Simple Maths books available in hardback
Boxes
Circles
Games (Number Games in paperback edition)
Number Puzzles
Printing
Railways

Simple Maths books available in paperback
Facts and figures
Numbers
Number Games
Number Puzzles

First paperback edition published 1999
First published in hardback in 1993 by
A & C Black (Publishers) Ltd
35 Bedford Row, London, WC1R 4JH
© 1993 A & C Black (Publishers) Ltd

A CIP record for this book is available from the British Library.
ISBN 0-7136-5270-5

Acknowledgements
Edited by Barbara Taylor
Mathematics consultant Mike Spooner

The photographer, author and publishers would
like to thank the following people whose help
and co-operation made this book possible:
the staff and pupils of Kenmont Primary School,
Jane Salazar, Jane Tassie, the Early Learning Centre,
Lego UK Ltd.

Apart from any fair dealing for the purposes of research and private study,
or criticism or review, as permitted under the Copyright, Designs and Patents
Act, 1988, this publication may be reproduced, stored, transmitted, in any form
or by any means, only with the express permission in writing of the publishers,
or in the case of reprographic reproduction in accordance with the terms of licenses
issued by the Copyright Licensing Agency. Inquiries concerning reproduction
outside those terms should be sent to the publishers at the above named address.

Typeset by Rowland Phototypesetting Ltd,
Bury St Edmunds, Suffolk.
Printed and bound in Italy by L.E.G.O. Spa

NEWCASTLE UPON TYNE CITY LIBRARIES	
Class No.	Acc No. C3 398234 C4 280708 00 93
Checked SJW	Issued 2/00

Newcastle
City Council

Education and Libraries Directorate
Newcastle Libraries & Information Service

Please return or renew this item by the last date shown. Books can be renewed at the library, by post or by telephone if not reserved by another reader.

Due for return	Due for return	Due for return
PLAYGROUP		

E4331/144

Number Games

Rose Griffiths
Photographs by Peter Millard

A & C Black · London

We can play lots of different card games.

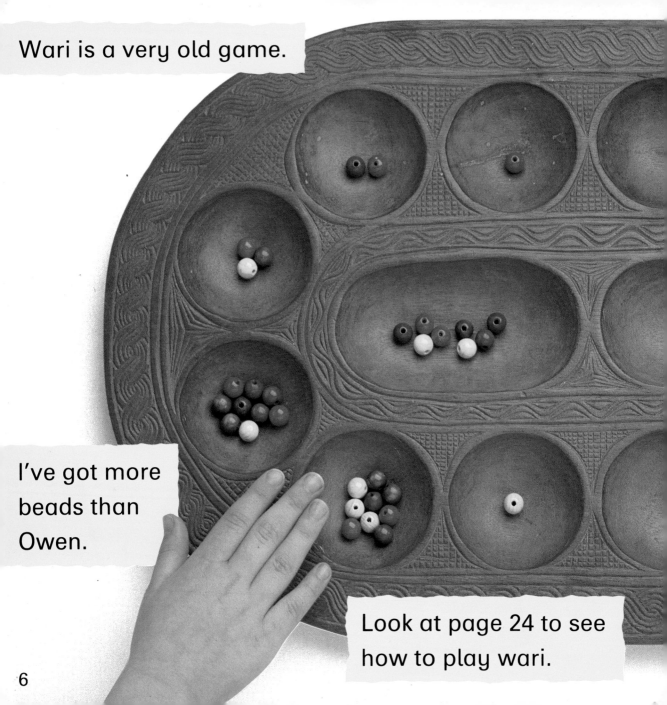

Wari is a very old game.

I've got more beads than Owen.

Look at page 24 to see how to play wari.

Computer games are new.

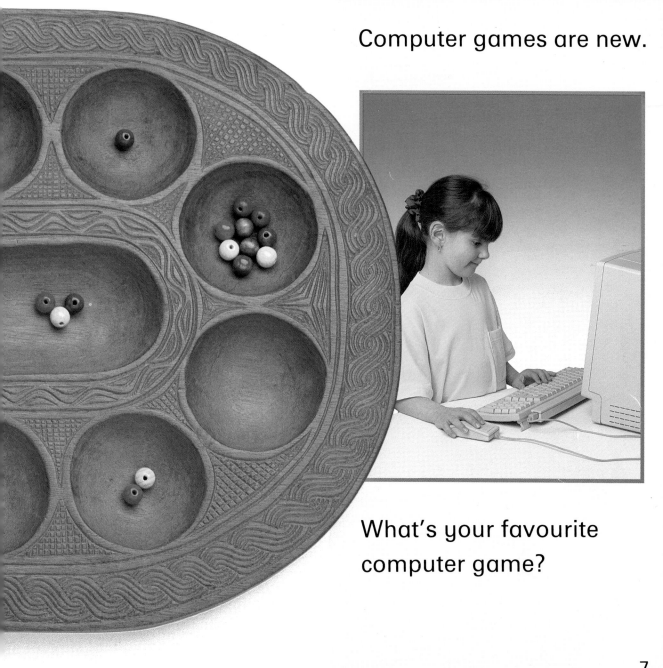

What's your favourite computer game?

We've just made up this game.

How do you think we play it?

How many animals will the farmer catch?

Owen drew this track for a racing game.

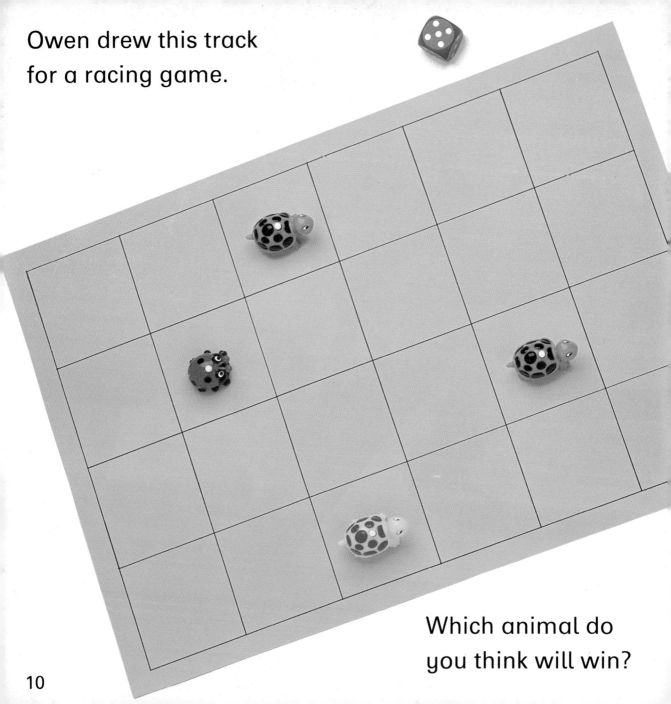

Which animal do you think will win?

He changes the rules sometimes.

Will the people catch the chickens?

Where does this number go?

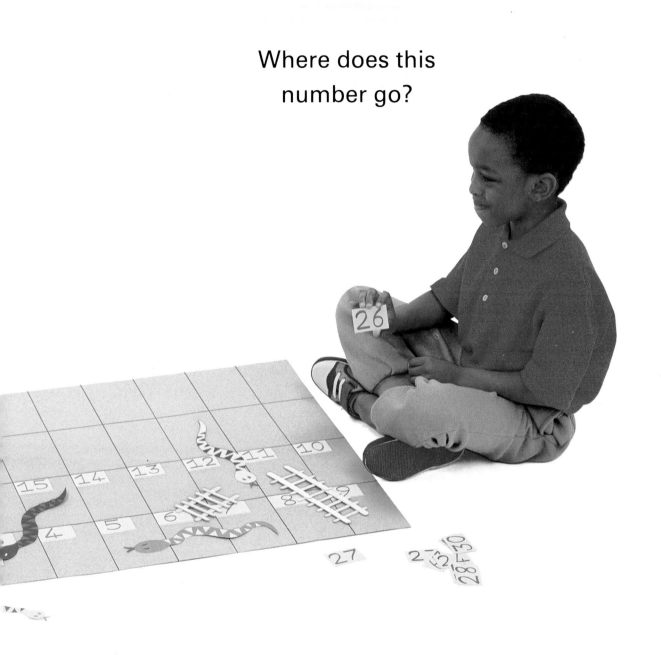

13

But some games just depend on luck.

Heads to move the green car. Tails to move the yellow car.

Dominoes depends on luck and skill.

Can you remember everything that's here?

What have I taken away?

Can you get better at this game?

I want to be first to match all the pictures on my card.

Does this game need luck or skill?

We're playing a stepping game.

If the spinner stops on red, Owen moves.

If it stops on blue, Alice moves.

Do we both have the same chance of winning?

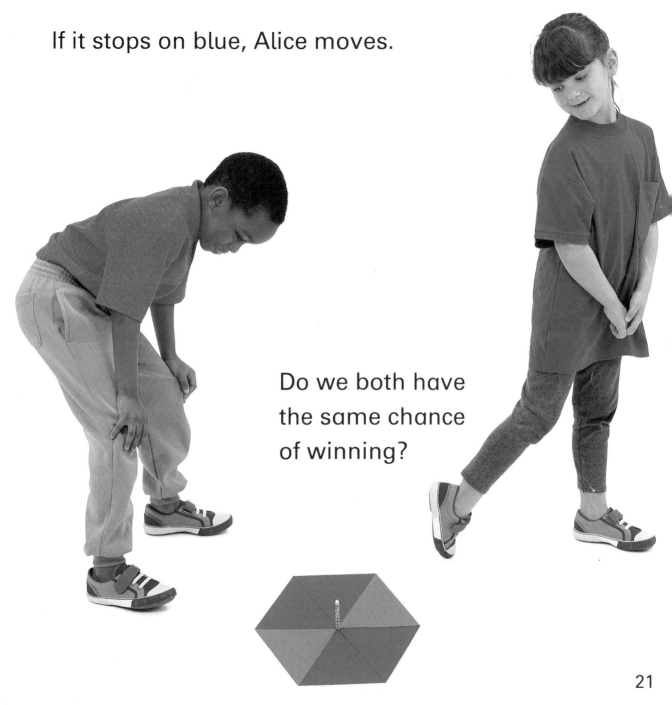

21

We like playing football.

Which games do you like playing?

More things to do

1. Big games
To make a giant race game, chalk a track on a playground and use your friends as the counters. Or try making a big game of noughts and crosses to play on a carpet. Use four long pieces of ribbon as the lines and five paper plates as the noughts. Cut up five more paper plates to make the crosses.

2. Play wari
Collect forty-eight beans, beads or smooth stones and make a board from two egg boxes and two margarine tubs. The two players each have a row of six holes and one hole to store the beads they collect during the game.

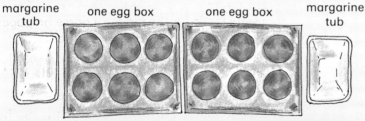

margarine tub | one egg box | one egg box | margarine tub

Here is one way to play. To start, put four beads in each of the twelve holes. When it is your turn, pick up all the beads in one of the six holes on your side. Then put the beads, one in each hole, in the next few holes going anti-clockwise around the board. If the last hole you put a bead in now has either two or three beads in it – and is not on your side – you can put the beads in your store. You can also keep the beads from the hole before the last one if this has two or three beads in it. But if you take the other player's last beads you lose the game! The game stops when one player has no beads left. If you have the most beads in your store, you win.

Find the page

This list shows you where to find some of the ideas in this book.

Pages 2, 3, 4, 5, 6, 9, 11, 15, 16, 20, 21
How to play games

Pages 4, 5
Card games

Pages 6, 7
Old and new games

Pages 8, 9, 10, 11, 12, 13, 15
Making up games

Pages 14, 15, 16, 17, 18, 19
Luck or skill?

Pages 17, 18
Memory game

Page 19
Matching game

Pages 20, 21
Fair and unfair games

Pages 22, 23
Favourite games

Notes for parents and teachers

As you read this book with children, these notes will help you to explain the mathematical ideas behind the different activities.

Games have rules Pages 2, 3, 4, 5, 6, 9, 11, 15, 16, 20, 21

Explaining and discussing the rules of a simple game requires children to think logically and to consider alternative ways of explaining things.

Sometimes, as with cards and dominoes, the same game pieces can be used, with different rules, to play several different games. People can alter games if they wish, but they have to agree on the rules they will use before they play a game. For example, they might change the rules about how to take turns or how to score points.

Old and new games Pages 6, 7

The game of wari (also called ayo, songe, congklak or mancala) has been played all over the world for hundreds of years. It is particularly popular in Africa and South-east Asia. Wari gives children practice in counting and thinking ahead. The rules of wari vary from one country to another.

Computer games are a relatively new invention. Like other games, they vary a great deal in how mathematically useful they are.

Inventing games Pages 8, 9, 10, 11, 12, 13, 15

Making and playing games can help children practise many mathematical skills, such as estimating, measuring, counting, matching, sorting and using shape. Children whose skills at measuring and drawing are still at an early stage may need an adult to draw simple tracks or boards for them. This leaves the children free to experiment with movable components until they are happy with the games they invent. Encourage children to change the rules and develop several versions of one game.

Chance or skill? Pages 14, 15, 16, 17, 18, 19

Ideas about chance take a long time to develop. For example, children may be convinced that you can learn to throw dice to get a particular number when you want it, or that wishing hard will help you. They need to repeat the same game several times and talk about their findings, so that they can begin to come to more realistic conclusions.

Many children learn most about strategies and tactics if they act as 'helper' to an older child or an adult player. This gives them the opportunity to discuss the next move. One useful tactic for the memory game on pages 17/18 is to sort the objects into groups that have something in common – for example, the scissors, spoon and paper clip are all metal.

Fair or unfair? Pages 20, 21

It's a good idea to let the children try out the spinning game for themselves. Since the spinner has four sections coloured blue and only two red ones, it is twice as likely that the child in blue will win. Playing a game like this helps children to think about probability.

Surveys Pages 22, 23

Children may enjoy carrying out a survey of favourite games. See how many different ways of sorting and classifying games the children can think of. Some possible categories might be: games of skill or chance; competitive or co-operative games; indoor games or outdoor games.